LOTS OF ROT

by Vicki Cobb   illustrated by Brian Schatell

J. B. Lippincott   New York

Also by Vicki Cobb

More Science Experiments You Can Eat
Science Experiments You Can Eat
Arts and Crafts You Can Eat
Magic...Naturally!

LOTS OF ROT

Text copyright © 1981 by Vicki Cobb

Illustrations copyright © 1981 by Brian Schatell

All rights reserved. Printed in the United States of America. No part of this book may be used or reproduced in any manner whatsoever without written permission except in the case of brief quotations embodied in critical articles and reviews. For information address J.B. Lippincott Junior Books, 10 East 53rd Street, New York, N.Y. 10022. Published simultaneously in Canada by Fitzhenry & Whiteside Limited, Toronto.

Library of Congress Cataloging in Publication Data

Cobb, Vicki.

Lots of rot.

SUMMARY: Discusses what causes rot and the role it plays in the cycle of living things and presents facts about mold, bacteria, and mildew. Includes experiments. 1. Microbiology—Juvenile literature. 2. Biodegradation—Experiments—Juvenile literature. [1. Biodegradation. 2. Microbiology] I. Schatell, Brian. II. Title.

QR57.C6 1981   576   80-8726

ISBN 0-397-31938-X   ISBN 0-397-31939-8 (lib. bdg.)

3   4   5   6   7   8   9   10

Contents

For Dara Gabrielle Zabb

1 ROTTEN STUFF

Want to smell something rotten? Take a deep breath by a garbage can. If it's rotten, your nose knows. All it takes is one sniff! Yukk!

Rotten stuff looks different from the way it starts out. An orange now has blue-green spots growing on it. A slice of bread may be covered with black powder or have small shiny bumps on it. A piece of celery once felt firm. Now it feels soft and mushy. Want to touch it? Yukk!

Food in your garbage is something sure to rot. But there are many other things that will also rot. Wood rots. So does leather, paper, cloth, and string. Almost anything that was once alive or once a part of a living thing can rot. It just takes time. Sometimes years.

Most people don't like rot. It spoils food and makes wood and cloth very weak. Rotten things can't be used the way people want to use them. So, long ago, people looked for ways to slow down rotting. They learned that rotting doesn't happen right away if you keep stuff cold or dry. Rotting happens fastest when things are warm and moist. But for thousands of years, no one knew what caused rot. Once we learned what caused rot, we found ways to stop it altogether.

The people who discovered the cause of rot were scientists. Scientists don't think rotten stuff is disgusting. They find it interesting. They understand the good side of rot.

Want to be a scientist? You, too, can discover what causes rot. This book will tell you how to explore the wonder and mystery of rot. It will tell you where to find rot. It will tell you about different kinds of rot. And it will tell you how to grow some rotten stuff of your very own.

Want to do some rotten stuff? Read on.

2 MOLDY FRUIT

Ever see a rotten lemon? Its skin has a coat of blue-green powder. It is no longer yellow. The blue-green powder is made up of millions of tiny plants that are too small to see with your eyes alone. Scientists did not know what these plants looked like until the microscope was invented, which was about 300 years ago. Without the microscope, we would not have discovered the true cause of rot.

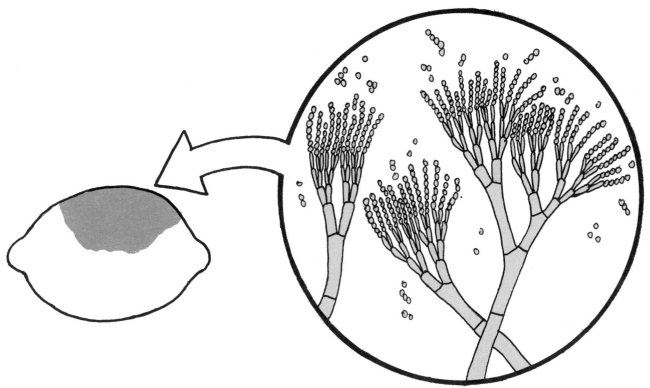

The tiny blue-green plants are called mold. Molds use lemon peel as food. As the mold grows, the lemon peel is used up. The mold gives off juices that make the lemon softer. The lemon will never be the same after it is used by the mold as food. Above is what a mold plant looks like under a microscope. Tiny threads grow into the lemon peel. They are like roots. Other threads grow into the air like stems. The brushlike ends of the mold contain millions of tiny blue-green balls. These balls are called spores. They are like seeds. New molds grow from spores.

Do an experiment to see how mold grows. Find out if mold on a lemon will grow on an orange.

You will need these things:

- a moldy lemon (ask the fruit man at the supermarket if you don't have one at home)
- a good orange
- a plastic bag
- a twist tie to close the plastic bag
- a magnifying glass

Put the moldy lemon in the plastic bag. Put the good orange in the bag next to the lemon. Make sure they are touching. Close the plastic bag with the twist tie.

Leave your experiment out on the counter. Watch what happens every day for a week. Use your magnifying glass for a close look. See if the orange catches the mold from the lemon. If it does, you know that the orange can be food for the mold. Do you think the mold can grow on other kinds of food? Try growing it on other fruits and on bread. Put moldy fruit in plastic bags next to bread, or an apple, or a piece of cake. Make up your own experiments.

3 CATCH A CASE OF MOLD

In the last experiment, the orange is touching the mold on the lemon. But you can get mold to grow without having the food touch the moldy fruit. Mold spores are always in the air. They float and they are carried everywhere by the wind. You can't see them but you can catch them. Here's how to do two experiments for catching mold.

You will need these things:

a slice of bread

a slice of left-over boiled potato

two jars with covers

a magnifying glass

Put the bread in one jar. Put the slice of potato in the other jar. Sprinkle the bread with water. Leave the jars open to the air for half an hour. Cover the jars. Put them in a closet. Take them out and look at them every day for a week. Use your magnifying glass.

Here's what to look for: Find white fuzzy spots. This is a different kind of mold from the mold that grows on oranges. The fuzz is made up of threads that grow into the food like roots and up in the air like stems. In a few days this fuzz will look black. That's because it's covered with tiny black spores. These spores fly off into the air. They are so small you don't know they are there. When they happen to land on bread or potato or some other food they grow into new mold. There are billions and billions of mold spores in the air. Most of them never get a chance to grow into mold.

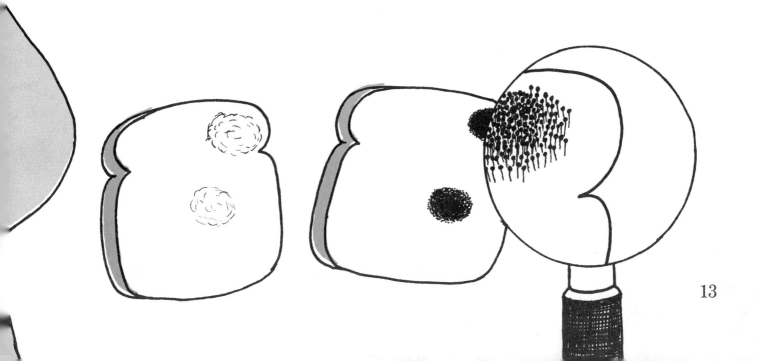

You may see blue-green spots. These are also a kind of mold. They are a cousin of the mold that grows on lemons.

You may see the same molds that grow on bread growing on the potato. You may also see shiny round white bumps on the potato. These bumps are made of millions of tiny plants that are much smaller than molds. They are also much simpler.

Here's how they look under the microscope: They may be shaped like round balls, or like rods, or like wavy worms. They do not have parts that look like roots or stems. They are called bacteria.

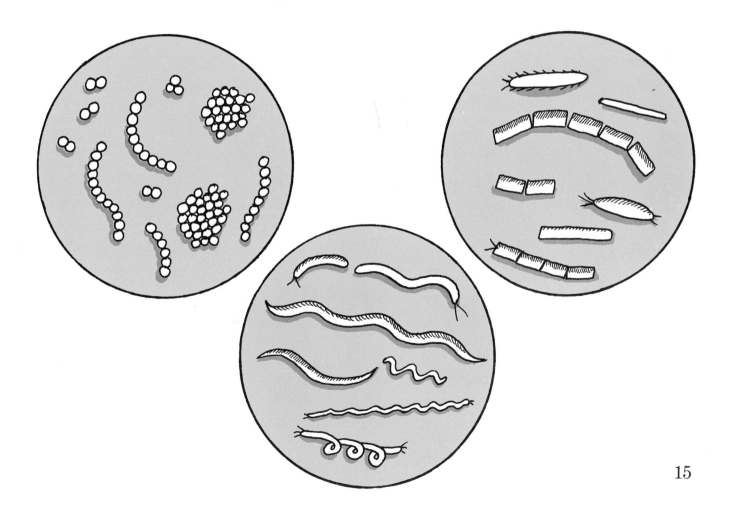

Bacteria are everywhere. They are carried by the air, like mold spores, ready to land somewhere and grow. They are in water and soil. Bacteria do most of the rotting of the world.

When scientists learned that living molds and bacteria cause rot, they worked to find ways to stop it. Heat kills molds and bacteria. Using heat to kill these tiny plants is called sterilization. If you sterilize food and then seal it off from the air, it will not get rotten. Canned food is protected from rot this way. Freezing also stops rot. Bacteria and molds don't grow on frozen food. But if frozen food warms up, it can start rotting.

4 ROTTEN COTTON

Mold can also grow on things you wouldn't want to eat. Mold can use cotton as food. You can grow a kind of mold called mildew on a cotton sock. Cotton cloth is made from cotton plants. Cotton plants grow fuzzy white balls. These balls have tiny plant threads that are woven into cloth. The threads that make up cloth are food for mildew.

You will need these things:

an old cotton sock that has lost its mate
a plastic bag
a twist tie

Wear the old sock for a day. Get it good and dirty and full of sweat from your foot. Find a deep puddle. Dip the sweaty sock in the puddle. You do all this to make sure invisible mildew spores get on the sock. The spores are in dirty places as well as in the air. Wring out the wet sock. Roll it up and stuff it in the plastic bag. Close the bag with a twist tie. Put it in a closet.

19

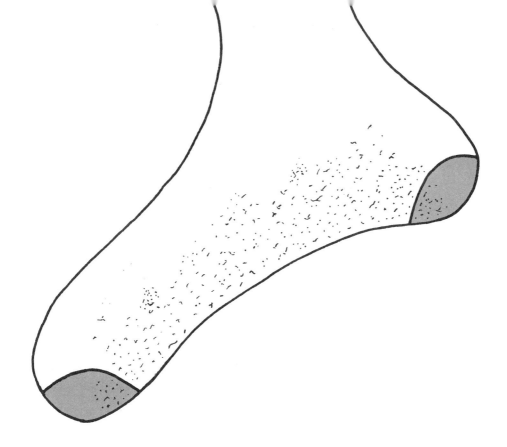

Check the sock every day for a week. Black specks will grow on it. These are mildew. Mildew grows where it's damp and where there is food like plant threads. Must you have a dirty sock to grow mildew? Do an experiment to find out. See if mildew will grow on a clean wet sock.

Mildew also seems to grow better in the dark. Do an experiment with one sock in the dark and one in the sunlight to see if this is so.

Go on a mildew hunt. Look at old clothes that have been stored in a basement. Books stored in damp places may also grow mildew, because paper is made from the same kind of plant threads as cotton.

Mildew smells musty. This is the stale smell of closed, damp rooms. If mildew grows for a long enough time, it makes cloth very weak. The cloth will fall apart when you pull on it.

Mildew can be a problem in your hamper. Don't put sweaty clothes or wet bathing suits in with the family wash! If the wash isn't done right away, you're asking for trouble! Wet clothes that sit in a hamper for a week can smell up the place. And they'll be ruined!

Mildew is not fussy about its food. Check for mildew

in the bathroom. Look in cracks in the tile and on the shower curtain. It grows on plastic, paint, leather, and soap film. See how many places you can find in your house where mildew grows. You'll find that mildew grows anyplace where it is damp and there is some kind of food.

5 DIRTY AND ROTTEN

Soil is full of all kinds of bacteria. A good way to rot something is to bury it. Of course, it takes a long time for soil bacteria to finish rotting something. But you can see the beginning of how soil bacteria rot stuff in the next experiment.

You will need these things:

    a large flower pot

    a spoon for digging

    soil from your yard or park

    toothpicks

    paper and pencil

    scissors

Fill the flower pot with soil.

Collect some things to bury.

Here are some ideas:

    a piece of apple or potato

    orange peel

    a piece of paper

    plastic wrap

    aluminum foil

    a lettuce leaf

Make signs for each object. Cut out a small piece of paper. Write the name of the object on the paper. Stick a toothpick through the end like a flag.

Bury each object in a big flowerpot. Stick a toothpick flag into the dirt over each buried object. Pour a glass of water over your experiment. Water it again in three or four days. Bacteria must have water to grow. Bacteria will not grow where it is dry.

Wait a week. Then dig everything up. Wash off the dirt if you can. It will stick where things are getting rotten. That is because bacteria are growing into the food. The soil sticks to the bacteria, and the bacteria stick to the food. Feel the rotting fruits and vegetables. They will be getting soft. If there is a peel, it will not be as rotten as the inside. Peels take longer to rot. They protect apples and potatoes from rot. Peels often have a coat of wax that keeps the bacteria from getting in.

Some things will not be rotten. They simply are not food for bacteria. Paper will rot, but it takes longer than a week. Plastic wrap and aluminum foil will not rot. They cannot be used as food.

Everything that rots is called biodegradable. Biodegradable is a long word made up of parts that have separate meanings:

"Bio" means "living."

"Degrade" means "to break down."

"Able" means "having the power to do something."

When you put the word together, biodegradable means "able to be broken down by living things." Biodegradable things end up broken down into crumbs. Rotting completely breaks them down. Litter collects when people toss away things that are not biodegradable, such as bottles, cans, and foil wraps.

If you like, you can bury everything in your experiment again. Dig it up after it rots for another week. Or leave it buried for a month or longer before you dig it up again.

6 COMPLETELY ROTTEN

The experiments in this book show how rotting starts. It takes a long time for rotting to be finished. You can see completely rotten stuff in the woods. Take your magnifying glass with you for a closer look.

Look at the leaves on the ground. The ones on top fell last fall. Some of them may be in pieces. These are starting to rot.

Dig under the top leaves with a stick. You will find leaves that are like crumbs. These are almost completely rotten. They fell years ago.

Find a rock and lift it up. Under the rock you may see worms. Worms eat dead plants and help break them down into crumbs.

Look for a dead log lying on the ground. It may have mushroomlike plants growing on it. They also help rot stuff. If it is rotten, you can break it apart when you kick it. Rotten wood is soft. There may be insects living in it. They help rot things, too. When you find crumbly stuff in the woods, you can be sure it has been rotting.

But if it's been there a long time, you can't always name what it was before it started rotting.

Rotting is very important for all living things. Suppose there were no molds, or bacteria, or insects, or worms to rot stuff. Falling leaves would hang around forever. They would pile high enough to cover the tree tops. Dead branches would pile up. Garbage would pile up. Soon the earth would be covered. There would be no room to live.

Dig into the forest floor. Grab a handful of the damp, black, crumbly stuff under the leaves. Smell it. You are smelling something completely rotten. It smells sweet and rich. Completely rotten stuff makes the soil rich. It fertilizes the soil so that new plants can grow in this rich soil. The end of rot becomes a part of new living things that will grow, then die and rot again. Rotting is part of the circle of living things.

WHICH LITTLE ROTTERS DID YOU GROW?

NAKED EYE   MAGNIFYING GLASS*   MICROSCOPE

BLUE MOLD ROT

Scientific name: *Penicillium*

Appearance: powdery blue-green

Favorite foods: cheese (blue cheese), bread, lemons, peaches, and other fruits

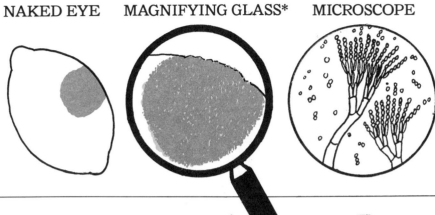

GRAY MOLD ROT

Scientific name: *Botrytis cinera*

Appearance: gray fuzz

Favorite foods: grapes, strawberries, and many other fruits and vegetables

RHIZOPUS SOFT ROT (also known as black bread mold)

Scientific name: *Rhizopus stolonifer*

Appearance: cottony with black dots

Favorite foods: bread and cake, vegetables, fruits

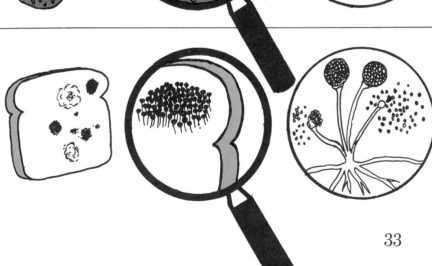

33

BLACK MOLD ROT (also known as "smut")

Scientific name: *Aspergillus niger*

Appearance: powdery, usually black but may also be brown or green

Favorite foods: vegetables (onions and tomatoes), bread, fruit, very sweet stuff such as jellies and jams, and very salty foods such as bacon

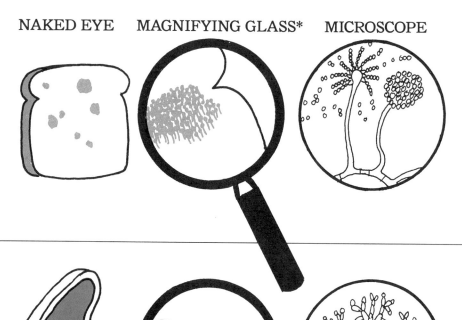

NAKED EYE   MAGNIFYING GLASS*   MICROSCOPE

GREEN MOLD ROT

Scientific name: *Cladosporium herbarum*

Appearance: thick, velvety, dark olive-green spots

Favorite foods: fruits and dark spots on beef

MILDEW

Scientific name: *Ascomycetes*

Appearance: black powdery spots on light materials, light spots on dark materials

Favorite foods: cloth, leather, paper, many plastics, paint, and soap film

*Magnifying glass is not here because it doesn't show a big enough difference.

BACTERIAL SOFT ROT

Scientific name: *Erwinia caroto-vora*

Appearance: mushy, water-soaked, sometimes bad smelling; sometimes seen as pink to reddish spots on potato slices

Favorite foods: vegetables and just about any other dead stuff that's juicy

NAKED EYE   MAGNIFYING GLASS*   MICROSCOPE

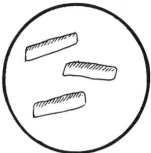

BACTERIA COLONIES

Scientific name: *Bacilli*

Appearance: roundish spots. Some may be raised and some flat. Shiny and smooth or rough. Color ranges from white to cream to brown.

Favorite foods: Bacilli grow on most foods but they need more water than molds

Bacilli do most of the world's rotting. They grow especially well where it is warm and moist.

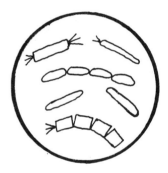

*Magnifying glass is not here because it doesn't show a big enough difference.